Top Techniques to Improve Concentration

Dean Thomas

Published by Books for Minds, 2023.

TOP TECHNIQUES TO IMPROVE CONCENTRATION

First edition. April 7, 2023.

ISBN: 979-8215529485

Written by Dean Thomas.

Table of Contents

Top Techniques to Improve Concentration

Step-by-Step Guide to Clearer Focus and Better Results

Welcome

Are you struggling to focus on your work? Does your mind wander off every few minutes? Do you find it hard to stay on task? If yes, then you're not alone. Concentration is a skill that takes time and effort to develop, and unfortunately, many of us haven't given it the attention it deserves. In this book, we answer the most frequently asked questions about concentration, including what it is, why it's important, and how you can improve it. Whether you're a student looking to ace your exams, or a professional trying to meet your deadlines, this book is all you need to know about concentration. So, sit back, relax, and let's dive in!

Disclaimer Notice:

The information contained in this book is for general information purposes only. While we endeavor to keep the information up to date and correct, we make no representations or warranties of any kind, express or implied, about the completeness, accuracy, reliability, suitability, or availability with respect to the book or the information, products, services, or related graphics contained in the book for any purpose.

Introduction

How to use the book effectively

1. Read the book thoroughly: The first step to using this book effectively is to read it thoroughly. Make sure you understand the concepts and techniques discussed in the book.
2. Identify your specific concentration issues: Identify what specific issues you have with concentration. This could be difficulty maintaining focus on a task, getting easily distracted, or having a short attention span.
3. Highlight key points and exercises: As you read the book, highlight key points and exercises that you think will be particularly helpful in improving your concentration.
4. Create a plan: Once you have identified specific issues with concentration and highlighted key points and exercises in the book, create a plan of action. Determine which exercises you will incorporate into your daily routine and how often you will do them.
5. Practice regularly: Practice the exercises regularly to see improvements in your concentration. Remember that it may take time and consistent effort to see significant results.
6. Monitor progress: Keep track of your progress over time. This will help you see the improvements you are making and identify where there is still room for improvement.
7. Apply the techniques to real-life situations: Finally, apply the techniques and strategies discussed in the book to real-life situations. Use them to help you stay focused and productive in your daily tasks and activities.

Frequently Asked Questions About Concentration

1. What is concentration?
2. Why is concentration important?
3. What are some tips for staying focused?
4. How do distractions affect concentration?
5. How can you improve your concentration?
6. How can I reduce distractions while studying?
7. What are some exercises to improve concentration?
8. What are the common problems with concentration?
9. How long should I sit and study without taking a break?
10. How can you maintain concentration for longer periods?
11. How can you use technology to improve concentration?
12. Are there any exercises or activities that can help improve concentration?
13. Are there any specific foods or supplements that can help improve concentration?
14. What are some effective study techniques for maintaining focus and concentration?

You will find very satisfying answers to the above questions within this book.

Chapter 1

Importance of concentration in learning and daily life

Concentration is essential for accomplishing any task effectively. In learning, concentration is critical as it determines how much information one can retain and understand. With good concentration, learning becomes more efficient, and an individual is better able to apprehend the subject at hand.

In daily life, concentration plays a significant role in completing tasks effectively. When we concentrate on a task, we can complete it promptly, and the result is usually of better quality. For example, while driving, it's vital to concentrate to avoid an accident. Similarly, during a meeting or presentation at work, concentration ensures that a person comprehends the information and contributes meaningfully.

Concentration also enables one to complete multiple tasks simultaneously without compromising quality. It helps one to prioritize tasks and manage time better.

In conclusion, concentration is vital for both learning and daily life. It enhances efficiency, enables one to prioritize tasks effectively, and ensures that work is completed to a high standard. Therefore, improving concentration skills is beneficial for an individual's personal and professional growth.

Common distractions that affect pupils' concentration

1. Mobile phones or electronic devices
2. Talking or chattering by classmates

3. Noise from outside or nearby rooms
4. Hunger or thirst
5. Boredom or lack of interest in the subject
6. Physical discomforts like back pain or headache
7. Emotional stress or anxiety
8. Difficulty in understanding the lesson or tasks
9. Health issues like allergies or illness
10. Visual distractions like posters, decorations, or moving objects.

Chapter 2

Understanding Concentration

Definition of concentration

Good concentration refers to the ability to focus and sustain attention on a particular task or goal without being easily distracted or losing focus. It involves being able to shut out distractions, stay alert, avoid multitasking, and be fully present in the present moment. Good concentration enables individuals to think more clearly and efficiently, stay engaged and motivated, and achieve higher levels of productivity, performance, and success. Some key attributes of good concentration include a clear and calm mind, the ability to prioritize tasks, good time management skills, and the ability to stay organized and disciplined.

Some factors that may affect a person's ability to concentrate include their level of interest in a task, their motivation, their physical and mental state, and the environment in which they are working. Good concentration is often considered an important skill for academic, professional, and personal success.

The science behind concentration

Concentration involves the ability to focus on one task or thought while ignoring other distractions. It is important for a variety of activities, including studying, working, and even daily life. Here are some key scientific factors that play a role in concentration:

1. Brain Waves: Concentration is related to the different types of brain waves that are produced by the brain. **Beta waves**, which are associated with increased alertness, attention, and concentration, are the most important brain waves involved in concentration.
2. Neurotransmitters: Neurotransmitters play a crucial role in concentration. **Dopamine**, for example, is a key neurotransmitter that helps to control motivation and attention. Low levels of dopamine can lead to low concentration and motivation.
3. Environment: The environment around us also plays a significant role in concentration. High levels of noise or interruptions can disrupt concentration and make it difficult to focus on a task.
4. Mental health: Depression, anxiety, and stress can all have an impact on concentration. These conditions can cause difficulties in focusing and maintaining attention.
5. Sleep: Lack of sleep can also have a negative impact on concentration. During sleep, the brain processes and consolidates information from the day, and this processing is important for concentration the next day.

Overall, concentration is a complex process that involves various factors in the brain and body. By understanding these factors, we can take steps to improve our concentration and optimize our performance in various tasks.

Factors that affect concentration

1. Distractions: External distractions like **noise, interruptions, and social media notifications** can divert your attention and make it difficult to concentrate.
2. Stress and Anxiety: Stress and anxiety can affect your concentration as when you are stressed or anxious, your mind is often racing with thoughts, and it can be difficult to focus on a particular task.
3. Lack of Sleep: Lack of sleep can affect your concentration, as well as make you less alert and less focused.

4. Hunger and Thirst: Hunger and thirst can also affect your concentration, as your body needs energy and hydration to function properly.
5. Health Conditions: Health conditions like **ADHD** (Attention deficit hyperactivity disorder), depression, and anxiety can affect your concentration.
6. Boredom: Boredom can make it difficult to concentrate, as your mind may wander off into other thoughts or activities.
7. Noise: **High levels** of noise can affect your concentration, as it can be difficult to focus on a particular task.
8. Multitasking: Trying to do multiple things at once can affect your concentration, as you may find it difficult to give your full attention to any one task.
9. Lack of Motivation: Lack of motivation can affect your concentration, as you may not be fully engaged in the task at hand.
10. Use of Drugs and Alcohol: The use of drugs and alcohol can affect your concentration, as they can alter your cognitive abilities and impair your ability to focus.

The different types of concentration

1. Focused concentration - The ability to direct your attention and focus on a specific task or activity for an extended period of time.
2. Selective concentration - The ability to pay attention to specific information while ignoring distractions or irrelevant information.
3. Sustained concentration - The ability to sustain attention over a prolonged period of time, especially during lengthy or repetitive tasks.
4. Spatial concentration - The ability to concentrate on spatial relationships or connections between objects, positions, or movements.
5. Divided concentration - The ability to multitask or divide your attention between multiple tasks or activities at the same time.
6. Creative concentration - The ability to concentrate on creative or imaginative tasks, such as writing, problem-solving, or brainstorming.

7. Open concentration - The ability to be open and receptive to new ideas, thoughts, and sensations without judgment or preconceptions.
8. Internal concentration - The ability to focus on your own thoughts, emotions, bodily sensations, or inner experiences, often through meditation or mindfulness techniques.

Chapter 3

The Benefits of Concentration

Success in academic performance

Concentration is an essential component of academic performance. It creates and sustains the attention spans that students require to attain good grades and perform well in school.

<u>Here are several ways concentration can help with success in academic performance:</u>

1. Improved Memory: Concentration aids in strengthening memory, leading to better long-term retention of information.
2. Better Time Management: With greater concentration, students can more ably focus on their studies, requiring less time to complete their assignments.
3. Reduced Stress: When students concentrate better, their stress levels can diminish because they feel more comfortable and confident in performing their academic tasks.
4. Increased Productivity: By concentrating intently on a specific subject, **students can learn more in less time**, thus increasing their productivity and performance.
5. Better Decision-Making: With greater concentration, students can analyze and process information more effectively, leading to better decision-making abilities.

In conclusion, the **ability to concentrate is crucial for academic success.** By staying focused and avoiding distractions, **students can improve**

their memory, time management, productivity, stress levels, and decision-making abilities, leading to better academic performance.

Better memory retention

Here are some tips that may help improve memory retention:

1. Make associations: Link new information to something you already know to help your brain retain it. For example, associate a new name with something you already know about that person.
2. Repeat information: Repeating information helps move it from short-term to long-term memory. Try to recall the information several times to solidify it in your memory.
3. Keep learning: Challenge yourself with new information regularly. Continually learning and exploring new subjects can help strengthen your memory.
4. Sleep: Sleep is crucial for memory retention. Consolidation of memories occurs during sleep, so make sure you get enough restful sleep.
5. Use mnemonic devices: **Mnemonic devices** help you remember information by creating associations between words or concepts. For example, the "ROY G. BIV" acronym helps people remember the colors of the rainbow.
6. Take breaks: Breaks can help prevent mental fatigue and improve memory retention. Take regular **breaks or switch to a different task** to avoid overloading your brain.

Enhanced creativity and problem-solving skills

Concentration plays a vital role in enhancing creativity and problem-solving skills. Here are some ways how concentration can help:

1. Focused Attention: Concentration helps to improve your focus and attention, which in turn allows you to pay more attention to details, leading to ideation and creativity. It aids in **filtering out distractions and extraneous thoughts,** leading to clarity of thought process.

2. Visualization: Concentration helps in **visualizing scenarios and examining a situation** from different angles. This can lead to discovering new solutions or ideas in problem-solving exercises.
3. Increased Knowledge Retention: Concentration allows for better retention of knowledge, leading to better recall and the ability to merge different concepts that lead to creative problem-solving.
4. Mindfulness: Concentration allows for improved **self-awareness and mindfulness**, leading to better problem-solving skills. When we concentrate, we are more present and less likely to engage in impulsive reactions.
In conclusion, concentration helps to **improve cognitive abilities** that can enhance creativity and improve problem-solving skills. It allows us to experience the moment, become more focused, **understand the problem clearly**, and develop fresh and innovative ideas.

Greater productivity and efficiency

1. Focused Work: Concentration means that you are able to focus your attention and energy on a specific task or goal without getting distracted. This helps you get **more work done in less time**, leading to greater productivity.
2. Increased Quality: When you concentrate on your work, you are **less likely to make mistakes** or produce poor-quality outputs. With focus and attention, you can produce **high-quality work consistently**, improving the efficiency and effectiveness of your work.
3. Time Management: Concentration enables you to manage your time effectively, helping you prioritize tasks and work on the most important ones first. This increases your efficiency, as you are able to spend your time on tasks that matter the most, reducing the time spent on tasks with lower importance.
4. Efficient Resource Utilization: Concentrated work helps you to use your resources effectively. You are able to identify and utilize your skills, knowledge, and expertise to their fullest potential, maximizing your productivity within the given resources.

5. Improved Decision Making: Concentration enables you to think critically and take logical decisions in a timely manner. This helps you to avoid unnecessary delays or errors, optimizing your performance and efficiency.

Overall, concentration ensures that you **work smarter, not harder**, leading to improved productivity, efficiency, and quality of work.

Improved mental and emotional well-being

Concentration can have a significant impact on our mental and emotional well-being. Here are some ways in which it can improve:

1. Reduce stress and anxiety: Concentration can help to calm our minds and reduce stress and anxiety levels. It allows us to focus on the present moment and let go of worries and distracting thoughts.
2. Enhance cognitive function: Concentration helps to improve cognitive function by increasing focus, memory, and attention span. It can help us learn and process new information more efficiently.
3. Boost self-esteem: When we concentrate and achieve our goals, it can boost our self-esteem and confidence. We feel a sense of accomplishment and pride in our work.
4. Improve relationships: Concentration can help us communicate more effectively with others. When we concentrate on what others are saying, we can respond more thoughtfully and empathetically, leading to better relationships.
5. Increase productivity: When we concentrate, we can complete tasks more efficiently and effectively. This leads to increased productivity and a sense of satisfaction.

Overall, concentration is an essential tool for improving our mental and emotional well-being. It helps us manage stress, process information more effectively, and improve our relationships with others.

Improved mental and emotional well-being refers to a state of being mentally and emotionally healthy. This means that a person can cope with everyday stressors and enjoys a positive sense of well-being. Improved mental and emotional well-being is characterized by a range of positive attributes such as happiness, life satisfaction, self-esteem, and optimism. People who have good mental and emotional health are better equipped to handle challenges and maintain healthy relationships, which consequently leads to a better quality of life. Some habits that can help improve mental and emotional well-being include regular exercise, adequate sleep, healthy eating habits, positive self-talk, spending time outdoors, and engaging in activities that bring joy and relaxation.

Chapter 4

Concentration Techniques

Technique 1: Optimizing your Study Environment

The importance of a conducive study environment

A conducive study environment is essential for students to achieve academic success. It refers to the **physical, social, and mental conditions** in which learning takes place.

Here are some reasons why a conducive study environment is important:

1. Reduces distractions: A **peaceful and quiet study** environment will help students concentrate, minimize distractions and be more productive.
2. Enhances learning: A good study environment helps students to **focus and process information faster**, leading to **better retention and comprehension** of the materials studied.
3. Creates motivation: A positive, inspiring study environment can motivate a student to be **more interested and enthusiastic** about their studies.
4. Reduces stress: A comfortable environment can **reduce anxiety, lower stress levels, and increase productivity and efficiency in learning**.
5. Boosts creativity: A conducive study environment can help **inspire creativity** and provide space for students to express themselves.

6. Encourages organization: An organized study environment inspires **good study habits** and can help students **stay on top of their work** and **avoid procrastination**.

Overall, a conducive study environment **is crucial in enabling students to perform** at their best academically, to study effectively, and to achieve their educational goals.

Tips for creating an ideal study environment

1. Find a quiet and comfortable place to study - Choose a quiet place where you are **unlikely** to be disturbed. A comfortable chair and desk are also essential for long study sessions.
2. Keep distractions to a minimum - Remove any objects or items that can distract you from your studies. This includes **closing unnecessary tabs on your computer or putting away any distractions like your phone, TV, or video games console**.
3. Make sure there is adequate lighting - Study in a **well-lit room to reduce eye strain and fatigue**. Use bright or **natural lighting** over dim, yellow light.
4. Stay organized - Keep your study materials **organized and within reach**. This includes pens, pencils, highlighters, textbooks, and any other materials you may need.
5. Take breaks - Take regular breaks to **stretch your legs**, get some **fresh air**, or have a **snack**. This helps **boost your energy and ability** to concentrate.
6. Find motivation - Surround yourself with motivating quotes, pictures, and images that can help **boost your morale and increase your focus**.
7. Have some snacks nearby - Keep a **healthy snack** or two nearby, as it can help **reduce hunger pangs** and keep you going.
8. Drink water – Proper hydration is essential for maintaining focus and concentration while studying.
9. Put on some instrumental music - Some **instrumental music** might help you concentrate more and get rid of any **background noise**.
10. Set goals - Setting **small and achievable goals** can help you stay motivated while studying and keep your progress steady.

How to minimize distractions and improve focus while studying

Here are a few tips for minimizing distractions and improving focus while studying:

1. Choose a **quiet and comfortable place** to study, away from noise and interruptions.
2. Remove any distracting items from your workspace, such as **your phone, gaming console, or TV**.
3. Set a specific goal for your study session and break it into **smaller, manageable tasks.**
4. Use the Pomodoro Technique, which involves **studying for 25 minutes** and taking a **5-minute break**, to improve focus and concentration.
5. Use headphones and listen to instrumental or **soothing music to block out background noise**.
6. Make a to-do list for the day and **prioritize your tasks** to minimize distractions and increase productivity.
7. Stay hydrated and take **regular breaks** to avoid burnout.
8. Get **enough sleep** and exercise regularly to improve concentration and focus.

By following these tips, you can minimize distractions and improve your focus while studying, thus increasing your productivity, and achieving better results.

Chapter 5

Technique 2: Effective Time Management

The value of managing time effectively

Managing time effectively has many benefits such as:

1. Increased productivity: When you manage your time well, you **get more done in less time**. You focus on the essential tasks, **avoid procrastination**, and **minimize distractions**, which results in **greater productivity**.
2. Reduced stress: Poor time management can lead to stress and anxiety, as you struggle to meet deadlines and constantly feel overwhelmed. By managing your time effectively, you can **reduce stress levels,** which ultimately improves your overall health and well-being.
3. Better work-life balance: When you manage your time well, you can allocate time for work, family, hobbies, and other activities that are important to you. This leads to a better work-life balance, which in turn, helps **you feel more fulfilled and satisfied** with your life.
4. Improved decision-making: Effective time management requires you to prioritize tasks, **analyze your goals and objectives**, and make decisions about what to focus on. As you practice this skill, you improve your decision-making abilities, which can benefit you in all areas of your life.
5. Greater success: Time management is a critical skill **for success in any field** or industry. By managing your time effectively, you can **achieve your goals more efficiently, meet deadlines, and make progress** toward your long-term objectives.

How to set SMART goals and prioritize tasks

<u>Some tips on how to set SMART goals and prioritize tasks:</u>

1) Start by defining your long-term goal: What do you want to achieve? Be **specific and write it down**.

2) Break down your long-term goal into smaller, achievable goals: This will help you to **create milestones and stay motivated**.

3) Use the SMART technique to set goals: SMART stands for **Specific, Measurable, Achievable, Relevant, and Time-bound**. Make sure your goals meet these criteria.

4) Prioritize your goals: Identify which goals are **most important** and need to be accomplished first.

5) Create a plan of action: This will help you to track your progress and stay focused on your goals. Be sure to **include deadlines** and milestones.

6) Stay focused: Stay committed to your goals and **take action** on a daily basis to achieve them.

7) Re-evaluate your goals and priorities as needed: As your circumstances change, your priorities may shift. Regularly **re-evaluate your goals and priorities** to stay on track.

Time management tools and techniques

1. To-do lists – Writing down tasks and **prioritizing** them helps you stay on track.
2. Pomodoro Technique – Breaking your work down into **25-minute intervals with short breaks** in between can help you stay focused and productive.
3. Time blocking – **Blocking off** specific periods of **time for certain tasks** can help you stay accountable to your schedule.
4. Eisenhower matrix – **Categorizing tasks** based on urgency and importance can help you prioritize and delegate accordingly.
5. Calendar apps – Scheduling appointments and deadlines on a **digital calendar** can help you **stay organized and on track**.

6. Project management tools – Tools like Trello, Asana, or Monday.com can help you **manage tasks and keep track** of projects.
7. Time tracking apps – Tracking the time spent on tasks can **help you identify where your time is going and optimize** your schedule.
8. Email management – Setting aside **specific times** to check and respond to emails can help you avoid interruptions and distractions.
9. Mind mapping – Visualizing your tasks and projects with a **mind map** can help you see the big picture and break it down into **smaller, manageable parts.**
10. Goal-setting – Setting **specific, achievable goals** can help motivate you and keep you focused on what's important.

Strategies for overcoming procrastination

Some evidence-based strategies that might help:
1. Break tasks into smaller parts: It is easy to feel overwhelmed and put off tasks if they seem too big. Breaking them into smaller, **more manageable parts** can make them feel less daunting.
2. Create a schedule: **Scheduling** specific times to work on tasks can help **eliminate procrastination**. By **setting aside time** for the task and **sticking to it**, you can avoid distractions.
3. Use a timer: Setting a **timer to work** for a set amount of time can **help focus your energy on the task**, instead of letting your **mind wander**.
4. Start with the hardest task: Tackling the **hardest task first** can help you build momentum and reduce stress.
5. Eliminate distractions: Try to remove distractions while working on tasks. If possible, **turn off your phone or log out of social media or other time-consuming websites.**
6. Find accountability: Sharing your goals with someone else can help **hold you accountable** and give you the support you need to **overcome procrastination.**
7. Reward yourself: Celebrate small successes and milestones along the way. Give yourself **a treat or a break** that rewards your progress.

8. Identify why you are procrastinating: Figure out what is causing your **procrastination** and take steps to **overcome it**. Maybe you feel **overwhelmed, lack confidence**, or are **afraid of failure**. Once you identify this, you can create a **plan to tackle these issues.**

Chapter 6

Technique 3: Physical Exercise

How physical exercise influences concentration

Physical exercise can positively impact concentration and cognitive function in several ways:

1. Increased blood flow to the brain - physical exercise **increases blood flow to the brain**, which can enhance **cognition** and **improve concentration.**
2. Release of neurotransmitters - exercise causes the **release of neurotransmitters like dopamine**, which can improve **focus and attention.**
3. Reduced stress and anxiety - physical exercise can **reduce stress and anxiety**, which can improve **concentration and focus.**
4. Improved mood - exercise can **improve mood and motivation**, which may positively impact concentration.
5. Better sleep - regular exercise can **improve the quality of sleep**, which can improve **concentration** and **cognitive function** during waking hours.

Overall, physical exercise can improve concentration and cognitive function by improving blood flow and releasing neurotransmitters in the brain, reducing stress and anxiety, improving mood and motivation, and improving sleep quality.

ifferent types of exercise that could boost brain function

1. Aerobic exercise: This type of exercise includes activities such as **running, cycling, swimming, and dancing**. Aerobic exercise improves blood flow, which helps to deliver more oxygen and glucose to the brain.
2. Resistance training: Resistance training, also known as **weight lifting**, can improve brain function by increasing the amount of gray and white matter in the brain.
3. Yoga and meditation: These practices **can reduce stress and anxiety**, which can improve cognitive function and lower the risk of depression.
4. High-intensity interval training (HIIT): **HIIT** involves short bursts of intense exercise followed by periods of rest. This type of exercise has been shown to improve memory and cognitive function.
5. Tai Chi: This low-impact exercise combines deep breathing and slow movements to improve **balance and coordination**, which can reduce the risk of falls and **improve cognitive function**.
6. Dancing: Dancing not only provides a **cardiovascular** workout but also improves **coordination, balance, and memory**.
7. Mind-body exercises: Mind-body exercises such as **Pilates** and **Qigong** combine physical movement and mental focus to improve cognitive function and reduce stress.

Techniques for incorporating physical exercise into a daily routine

1. Start small and build up gradually: If you are not used to exercising, it can be easy to get discouraged by trying to do too much at once. Instead, **start with short periods** of exercise and gradually increase the duration and intensity over time.
2. Find activities you enjoy: If you don't like a particular type of exercise, it's unlikely that you will stick with it for very long. Try a **variety of different activities** until you find ones that you enjoy, whether it's yoga, swimming, cycling, or dancing.

3. Make it social: Exercising with **friends or family** members can make it more **fun and enjoyable**, and also helps to keep you **accountable**.
4. Prioritize exercise: Make physical activity a priority in your schedule by **blocking off time for it each day**. Treat it like an important appointment that can't be missed.
5. Incorporate exercise into your daily routine: Find ways to incorporate physical activity into your **daily routine**, such as taking the stairs instead of the elevator, walking or biking to work, or running errands on foot.
6. Make it a habit: Research suggests that it takes about **21 days** to form a new habit, so commit to making exercise a regular part of your daily routine for at least that long.
7. Mix it up: Try a variety of different types of exercise to keep things interesting and challenging. This will also help to **prevent boredom and burnout**.

Chapter 7

Technique 4: Healthy Eating and Hydration

The role of nutrition in boosting concentration

Nutrition plays a crucial role in maintaining good health, particularly cognitive health. In sufficient amounts, proper nutrients can help improve focus and concentration.

Foods that improve focus, memory, and overall mental health

1. Blueberries: Blueberries are high in **antioxidants** and compounds that help protect the brain from **oxidative stress**, which can damage **brain cells.**
2. Omega-3 Fatty Acids: These are essential components of brain cell membranes and help promote good **cognitive** function. Foods that are rich in omega-3 fatty acids include **oily fish, nuts, and seeds.**
3. Iron: A lack of iron in the diet can lead to anemia and decreased **cognitive** function. Foods that are high in iron include **lean meats, poultry, seafood, beans, tofu, and green leafy vegetables.**
4. B Vitamins: Vitamins B6, B12, and folic acid are essential for proper brain function and can help improve **concentration.** Foods that are high in B vitamins include whole **grains, meat, poultry, fish, beans, and dark green vegetables.**

5. Water: Dehydration can lead to poor cognitive function, including a decreased ability to **concentrate**. Drinking enough water can help maintain **focus and alertness** throughout the day.
6. Eggs: Eggs are a rich source of choline, a nutrient that is essential for **brain development** and **cognitive** function.
7. Dark chocolate: Dark chocolate is high in flavonoids, which can help improve **blood flow** to the brain and support **cognitive** function.
8. Turmeric: Turmeric contains curcumin, a compound that has been shown to improve **brain function** and protect against **cognitive** decline.
9. Avocado: Avocado is high in healthy fats that can help improve **brain function** and protect against **cognitive** decline.
10. Green tea: Green tea is high in antioxidants that can help protect the brain from damage and support **cognitive** function.

The importance of hydration and how it influences brain function

Hydration is **one of the most important factors** that influence brain function. The human brain is made up of about 75% water, and it needs a constant and sufficient supply of water to maintain its functions. Hydration also plays a crucial role in regulating body temperature and removing waste products, which indirectly affects **brain health.**

When we are dehydrated, our brain cells can't function normally, which can lead to symptoms such as **fatigue**, **headaches**, **confusion**, and **difficulty concentrating**. Even mild dehydration can impair **cognitive performance, memory, motor skills, attention, and mood**, making it more difficult to think, learn, and perform mental tasks effectively.

In contrast, drinking enough water can have a positive impact on **brain function, including memory, focus, and cognitive flexibility**. Research has shown that **drinking water before a test** or cognitive task can improve performance, especially for tasks that require **sustained attention or mental effort**.

Moreover, hydration status has been linked to brain structure and function, with low hydration levels being associated with **reduced brain volume** and cognitive decline in older adults. Adequate hydration has also been shown to protect against conditions like stroke, dementia, and Alzheimer's disease.

In summary, staying hydrated is crucial for maintaining optimal brain function and overall health. It is recommended that adults drink **at least 8 glasses of water per day**, and **more if they are engaging in physical activity**, living in hot or humid climates, or experiencing illness.

Strategies for incorporating healthy eating habits into your lifestyle

1. Plan ahead: Planning your meals ahead of time can help you make **healthy choices** and avoid impulsive decisions.
2. Cook your own meals: Cooking your own meals can help you control the **ingredients and portions**, and make healthier choices.
3. Keep healthy snacks on hand: Stock your pantry and fridge with healthy snacks such as **fruits, vegetables, nuts, and low-fat dairy products**.
4. Eat at regular intervals: Eating at **regular intervals** can help keep your metabolism running and prevent overeating.
5. Choose **healthy options** when eating out: Look for low-fat, low-calorie options on the menu, and avoid high-calorie, high-fat choices.
6. Practice portion control: Use smaller plates, measuring cups, and serving spoons to help **control your portions**.
7. Reduce your intake of processed foods: Try to eat fresh, whole foods instead of highly **processed or fast-food options**.
8. Limit your intake of **sugary drinks**: Opt for water, unsweetened tea, or low-fat milk instead of sugary drinks like soda or fruit juice.
9. Learn to read labels: Look at nutrition labels to **identify unhealthy ingredients** and make informed choices.

10. Be mindful of your eating habits: Pay attention to your **hunger and fullness signals**, and avoid eating when you're **bored, stressed, or emotional**.

Chapter 8

Technique 5: The Power of Breaks

The science behind taking breaks and its effect on concentration

Research has shown that taking regular breaks can have a **positive effect on concentration and productivity**. Our brains have a limited capacity for sustained attention, and taking breaks allows them to rest and recharge. Studies have also shown that taking breaks can help prevent burnout, reduce stress, and improve overall well-being. However, the frequency and duration of breaks may vary depending on the individual and the task at hand. It is important to find a balance that works best for you and your personal needs.

Strategies for planning breaks

1. Schedule breaks in advance: Plan your breaks throughout your day by **setting short intervals** at regular intervals during the day. Knowing ahead of time when your breaks are scheduled helps signify an end time to a task and allows for a break in focus.
2. Use a timer: Use a timer to keep track of your break intervals. Set a **timer** to go off every 60-90 minutes throughout the day to remind you to take a break.
3. Determine the length of the break: **Figure out** in advance how much time you need for each break. Five to ten minutes may be enough time

for a quick break, while a 20-30 minute break may work better for longer rests.

4. Plan activities: Determine ahead of time **how you want to spend** your break. If your workday is intense physical activity, rest or taking a short nap may be what you need. Alternatively, if your workday is filled with sitting, stretching, and movement breaks may be something to consider. Consider activities that allow you to **disconnect from work mentally** or bring negativity at bay.

5. No Distraction policy: Make breaks a sacred time that is **free of distraction**. During your breaks sign out from your phone, log out from your computer, or find a spot of solitude where you can be **quiet and introspective or do something that brings you joy**.

6. Prioritize your breaks: It's okay to take breaks, but you should **prioritize** your work ahead of unnecessary breaks. Take breaks when you need them, remember to keep moving forward with work, and do **not waste any valuable time**.

7. Identify the frequent points of distractions: Trace a pattern of how you waste your time and incorporate positive habits during those break moments. For instance, if in the morning, 10 am when you usually check **TikTok and Instagram**, allocate some time for the platforms during that break period.

8. Reward yourself: Associate work with rewards; When you take productive breaks or meet your work deadline, plan something to reward yourself, like **taking yourself out for lunch or buying something you enjoy**. It may **motivate** you to work harder, focus more, and continue to take healthy breaks.

How to incorporate breaks into a study schedule

<u>The Pomodoro technique.</u> This technique divides studying into **25-minute increments, each followed by a 5-minute break**. After **four Pomodoros**, you take a longer **15–20-minute break**. This method can help **maintain focus and motivation** by giving short, frequent periods

of rest that **don't disrupt your concentration.** However, you can **adjust the method to better suit your study style and preferences.**

Here are a few tips that can help you incorporate breaks into your study schedule:

1. Determine the best study time for you: Some people prefer studying in the morning, while others prefer nights. Whatever **your preference**, **identify the time of day** when you are most alert and productive.
2. Set study goals: Set clear and achievable goals for your study session. This will help keep you **focused and motivated** and make it easier to schedule breaks.
3. Choose the right break activities: During breaks, it's vital to engage in **activities that help you recharge**, like going for a **walk**, **stretching**, or even taking a **nap**. Avoid activities that stress or **drain you emotionally**, such as **social media or watching TV shows.**
4. Organize the break schedule: Once you've established your study goals, incorporate **regular breaks** into your schedule. Consider how often you need pauses based on your work period and goals. For instance, use the **Pomodoro Technique**, which is 25 minutes of studying and a 5-minute break.
5. Stick to your break schedule: During the breaks, try **not to check your phone, or emails, or engage in any activity that distracts you.** Stick to your break schedule, and once it's over, **get back to work**.

Chapter 9

Technique 6: Eliminating Distractions

Common distractions that affect pupils' concentration

1. Mobile phones: The use of mobile phones by pupils **during class** can be a significant distraction, especially if the phone rings or **buzzes** with notifications.
2. Social media: Pupils often get distracted by **social media platforms such as Facebook, Instagram, Twitter, etc. during class**.
3. Friends: Having friends sitting near them can be a **huge distraction** as pupils tend to talk to them, **play or joke** around with them **during class.**
4. Classroom environment: A noisy, cold, or uncomfortable **classroom environment** can also cause pupils to **lose focus**.
5. Lack of interest: If the topic being taught does not **interest** pupils, they may lose focus and become easily distracted.
6. Fatigue: Tired pupils who **lack sleep** or have had a long day may struggle to concentrate during class.
7. Hunger: Hungry pupils may get easily distracted by the feeling of an **empty stomach** or start to daydream about food.
8. Personal issues: Pupils who have **personal issues** such as family problems, relationship troubles, or financial difficulties may find it challenging to focus on classwork.

Techniques for minimizing distractions

1. Create a quiet and organized workspace: Set up your workspace in a **quiet, organized, and clutter-free area**. This can help reduce unnecessary stimuli and distractions, allowing you to **concentrate on your task**.
2. Set boundaries: Communicate to others that you need **uninterrupted** time to work. Set specific times when you are available and when you need to **be left alone**.
3. Turn off notifications: Turn **off notifications on your phone, email, and social media accounts**. Constant alerts can be interruptions and break your concentration.
4. Use time management techniques: Use techniques like the **Pomodoro technique**, where you break your work into time intervals, to help ensure that you are **working efficiently and productively**.
5. Take breaks: Taking regular breaks can help increase **productivity** and **reduce burnout**. Make sure you use your break time to **get away from your workspace** and recharge.
6. Practice mindfulness: Mindfulness practices, like meditation, can help train your brain to focus and **reduce the impact of distractions**.
7. Use productivity apps: There are several **productivity apps** that can help you stay focused and minimize distractions. Use them to **set goals, track progress, and visualize your progress**.

How to create boundaries and establish priorities

1. Assess your needs and desires. Determine which areas of your life require **boundaries** and **prioritization**. Consider your personal life, work life, relationships, and hobbies.
2. Identify limiting beliefs. Take note of any beliefs or **thoughts** that may be hindering you from creating boundaries or prioritizing your time.
3. Set goals. Envision what you want to achieve and **write down** specific, measurable goals to help you reach them.

4. Create a schedule. Plan out your week and **allocate time** for the things that matter to you most. Consider time for self-care, downtime, and hobbies.
5. Learn to say no. It's essential to **learn to say no** to requests or demands that don't align with your goals or priorities.
6. Communicate your boundaries. Let others know your limits and boundaries, so they **don't infringe on your time or energy**.
7. Stick to your priorities. Prioritize the things that matter most to you and **stay committed** to them. Avoid distractions or activities that don't align with your values or goals.

Chapter 10

Technique 7: Sleep and Rest

Why sleep and rest are essential for concentration

Based on research and studies, sleep and rest are crucial for concentration for the following reasons:

1. Improved cognitive function: Getting enough sleep and rest helps the **brain to function better**. When the brain is well-rested, it can process information more efficiently, which can enhance **concentration.**
2. Reduced stress and anxiety: Lack of sleep and rest can lead to stress and anxiety, which can interfere with concentration. Adequate sleep and rest can reduce **stress and anxiety levels**, making it easier to focus on tasks.
3. Increased creativity: Rest and sleep can help promote creativity. When the brain is well-rested, it can come up with more **creative ideas and solutions**.
4. Enhanced memory consolidation: Getting enough sleep and rest helps consolidate memories, making it **easier to remember** information and stay focused on tasks.
5. Reduced distractions: When the body is well-rested, it is **less likely to experience distractions and boredom**, which can interfere with concentration.

In conclusion, sleep and rest are essential for concentration because they improve cognitive function, reduce stress and anxiety, increase creativity, enhance memory consolidation, and reduce distractions. Therefore, it is important to prioritize sleep and rest in our daily routines to maintain optimal focus and concentration.

How to improve sleep quality and duration

Here are some tips to improve sleep quality and duration:

1. Stick to a consistent sleep schedule: Try to go to bed and wake up at a **similar time every day**, even on **weekends**.
2. Create a bedtime routine: Develop a **calming routine** before sleeping such as taking a warm bath, reading a book, or listening to soft music.
3. Create a relaxing **sleep environment**: Keep the bedroom dark, cool, and quiet. Use comfortable pillows and mattresses.
4. Avoid caffeine, alcohol, and nicotine: Intake of these **substances** can negatively affect the duration and **quality of sleep**.
5. Limit screen time: Exposure to blue light from electronic devices can affect sleep natural rhythms. Avoid using **electronic devices** at least 30 minutes before bedtime.
6. Regular exercise: Being **physically active** helps to improve sleep naturally.
7. Learn relaxation techniques: Use **relaxation techniques** such as meditation, deep breathing, or visualization to help unwind and sleep better.
8. Consult a healthcare professional: Consult a healthcare professional if you are experiencing **consistent sleep disturbances**.

Strategies for creating a sleep-friendly environment

Creating a sleep-friendly environment is essential to help you sleep better and wake up refreshed.

Here are some strategies that can make your bedroom the perfect sleep sanctuary:

1. Keep the room cool: The ideal temperature for sleeping is between 60-67°F. Make sure your **room** gets sufficient ventilation and install an air conditioner or use a fan to keep cool air circulating.

2. Invest in comfortable bedding: Choose sheets, pillows, and a mattress that are **comfortable** and promote good rest. Consider options with breathable materials, like cotton or linen, and a firmness level that suits your sleeping position.
3. Limit noise and light: Use blackout curtains or blinds to block out **outside light**, and reduce noise by using earplugs, a white noise machine or strategically placing soft furnishings around the room.
4. Declutter your space: A tidy bedroom is **conducive to relaxation** and comfortable sleep. Therefore, declutter your room of unnecessary items and create storage solutions for those essential items.
5. Designate your bedroom for sleep: Avoid watching **TV**, **reading, or working in bed**. You want to train your mind and body that the bedroom is a haven for sleep and relaxation.
6. Try calming colors: Soft, calming colors like blue, green, or lavender can create a **tranquil environment**, helping your mind and body relax.
By creating a welcoming sleep environment, you can improve **the quality of your rest** and help yourself wake up feeling more refreshed and rejuvenated each day.

Benefits of taking breaks and rest for concentration

Here are some benefits of taking breaks and resting for concentration:
1. Improves focus: Taking regular breaks and rest can help improve **focus and concentration** when performing complex or demanding tasks. This is because our brain gets tired after extended periods of focus, and taking a break helps to recharge and **rejuvenate** the mind.
2. Reduces stress: Taking breaks and rest can help **reduce stress levels** and improve overall mental well-being. When we take breaks, we allow our brains to relax, which can help reduce anxiety and tension.
3. Boosts productivity: Studies have shown that taking breaks can actually **boost productivity** by allowing our brains to **recharge and refocus**. This means that taking regular breaks and rest can help us get more done in less time.

4. Enhances creativity: Taking breaks and rest can also help enhance **creativity and problem-solving abilities**. This is because it allows the brain to explore new ideas and think outside the box.
5. Improves memory: Taking breaks and rest can also improve **memory** and **learning retention**. This is because our brains need time to **consolidate information** and taking breaks can help facilitate this process.
Overall, taking breaks and rest is essential for maintaining good mental health and improving cognitive abilities, including focus, productivity, creativity, memory, and learning retention.

Chapter 11

Technique 8: The Importance of Mindset

The significance of a growth mindset over a fixed mindset

A growth mindset is a belief that **intelligence**, **skills**, and **abilities** can be developed and improved through hard work, dedication, and persistence. A fixed mindset, on the other hand, is the belief that intelligence, skills, and abilities are predetermined and unchangeable.

Research has shown that individuals with a growth mindset are more likely to **embrace challenges**, **persist** through setbacks, and ultimately **achieve higher levels of success.** They are also more likely to take **constructive feedback** as an opportunity to learn and grow, rather than viewing it as a personal attack.

On the other hand, individuals with a fixed mindset are more likely to avoid challenges, give up easily, and view feedback as a threat to their sense of self-worth. This can lead to a **lack of motivation**, **low self-esteem**, and **limited personal development**.

In summary, having a growth mindset can lead to greater achievement, personal growth, and overall well-being, while a fixed mindset can hinder progress and limit potential.

How to develop a positive mindset and focus on the learning process

Some tips to help you develop a positive mindset and focus on the learning process:

1. Cultivate a Growth Mindset: A growth mindset is one where you believe that your **abilities** can be developed and improved through hard work and dedication. This means that no matter how **challenging** a task may seem, you approach it with the belief that you can ultimately learn and improve.
2. Practice Self-Compassion: Learning can be difficult and at times, you may make mistakes or struggle. It's important to be kind to yourself during these times and **acknowledge** that mistakes are a natural part of the learning process.
3. Set Realistic Goals: When setting goals, make sure they are **realistic** and **achievable**. This allows you to focus on making progress and meeting your goals, rather than becoming discouraged if you fail to meet **unrealistic** expectations.
4. Focus on the Process, Not the Outcome: Instead of solely focusing on the result or **outcome**, try to focus on the process or journey of learning. This can help you enjoy the learning experience and appreciate the **progress** you make along the way.
5. Embrace Challenges: Embrace challenges as an opportunity to **learn and grow**. Remember that challenging **experiences** may be uncomfortable now but can ultimately lead to **greater personal and professional growth**.
6. Practice Mindfulness: Take time to reflect on your **thoughts and feelings related to learning**. Mindfulness can help you stay present now, **improve your self-awareness**, and manage any **negative thoughts** or emotions that may arise.

By implementing some of these strategies, you can develop a positive mindset and focus on the learning process. Remember, learning is a lifelong journey, and with the right mindset, it can be an enjoyable and fulfilling experience.

The effects of positive self-talk on concentration

Positive self-talk refers to the **internal dialogue** or **thoughts** that individuals use to **reinforce** positive beliefs and attitudes about themselves. Concentration, on the other hand, refers to the ability to **focus and pay attention** to a **specific task or activity.** Both positive self-talk and concentration are **critical elements** that contribute to an individual's **overall performance, productivity, and success.**

Several studies have indicated that positive self-talk can significantly **enhance concentration** and **focus**. By using positive affirmations and **self-talk techniques**, individuals can improve their **self-confidence and self-esteem**, which can lead to better concentration and performance in various activities. Positive affirmations such as "**I can do this**," "**I am capable of achieving my goals**," and "**I am focused and determined**" can reduce negativity, anxiety, and stress that often hinder concentration.

Moreover, studies have found that positive self-talk can help individuals stay **motivated** even during **challenging tasks** or activities. Additionally, positive self-talk can enhance individuals' ability to **manage distractions** and **maintain focus** on their goals, leading to better **concentration and improved performance**.

In conclusion, positive self-talk can have significant effects on concentration, helping individuals stay motivated, manage distractions, and stay focused on their goals. Incorporating positive affirmations and self-talk techniques can enhance self-confidence, reduce stress and negativity, and improve overall concentration and performance.

Chapter 12

Technique 9: Simplicity and Minimalism

The benefits of minimalism and simplicity on concentration

Some information about the benefits of minimalism and simplicity of concentration:

1. Reduced Distractions: When you have **fewer things** around you, there are fewer distractions to take your attention away from your work. Minimalism allows you to focus on what's important and eliminates **non-essential visual clutter**.
2. Clearer Mind: Minimalism creates a sense of calm and **clarity** in your mind, which aids concentration. By decluttering your physical surroundings, you can also **declutter** your thoughts, making it easier to focus.
3. More Productive: When you have fewer things to worry about, you can focus more on the task at hand. With minimalism, you are more likely to **complete** tasks in a **shorter** amount of time, increasing productivity.
4. Better Health: Cluttered and chaotic environments can increase stress levels, which can negatively impact your health. Minimalism helps you to create a more **peaceful** and **serene** environment, lowering stress levels and promoting overall well-being.
5. Increased Creativity: Simplicity and minimalism can stimulate creativity. When your mind is free of distractions and clutter, you have more space to be imaginative and innovative. By creating a minimalist environ-

ment, you may be able to tap into your creative potential and improve concentration.

How to declutter and simplify your space

Here are some tips for how you can declutter and simplify your space:

1. Start small: Begin by decluttering one area of your space, such as a counter or a drawer. This will help build **momentum** and **encourage** you to tackle larger areas.
2. Identify what you need and what you don't: Go through your belongings and determine what items you **use and what you don't**. Consider donating or selling items that are no longer needed.
3. Consider functionality: Arrange your space in a way that is **functional** for you. Keep everyday items easily **accessible** and store items that are used **less frequently** in a safe space.
4. Create designated spaces: Establish designated areas for items such as keys, mail, and paperwork. This will help keep these areas **organized** and **prevent** clutter from spreading.
5. Regularly purge and maintain: Regularly go through your belongings and get rid of items that are no longer needed. Additionally, **take time each week** to maintain and clean your space.

By decluttering and simplifying your space, you can reduce stress and create a more peaceful home environment.

Strategies for reducing decision fatigue and mental clutter

1. Prioritize decisions: Start with the most important decisions that need to be made first and **work through the less important ones later**.
2. Limit decision-making: Avoid making **too many decisions** in a short amount of time. Try and **schedule** important decisions for a later time when you are more alert.

3. Take breaks: Take regular breaks throughout the day to recharge and avoid burnout. This can help **clear** your mind and enable you to make **better** decisions.
4. Simplify your choices: Reduce the number of decisions you need to make by simplifying your options, especially for **routine** or **mundane** tasks.
5. Avoid unnecessary stimuli: Limit distractions such as **social media or television** that can contribute to decision **fatigue and mental clutter**.
6. Create a routine: Set up a consistent routine for decision-making and problem-solving, so that you can **minimize** the need for making new decisions.
7. Practice mindfulness: Practice mindfulness meditation or relaxation techniques to clear your mind and reduce stress. This can help you make clearer decisions.
8. Delegate tasks: If possible, delegate tasks to others, so you have fewer tasks to prioritize and make decisions about.
9. Get enough sleep: Ensure you are getting adequate sleep, which allows your mind to relax and refresh, promoting good decision-making.
10. Stay organized: Keep your workspace and schedule organized, reducing the clutter to help clear your mind, and making it easier to make decisions.

Chapter 13

Case studies

1. Mindfulness-Based Stress Reduction (MBSR) in Medical Students
A study published in the Journal of Medical Education and Curricular Development explored the effectiveness of a mindfulness-based stress reduction program in medical students. The program involved sitting meditation, mindfulness of breath, body scan, and mindful movement. The study found that the program significantly reduced perceived levels of stress and improved mindfulness skills in medical students.
2. The Use of Biofeedback to Improve Concentration
A study published in the International Journal of Neuroscience evaluated the effectiveness of biofeedback in improving concentration. Participants received biofeedback training, which involved learning to regulate their brain waves through visual and auditory feedback. The study found that after receiving biofeedback training, participants showed significant improvements in concentration, attention, and overall cognitive performance.
3. Cognitive Behavioral Therapy (CBT) in Children with ADHD
A study published in the Journal of Attention Disorders explored the effectiveness of CBT in children with ADHD. The intervention involved training children to use cognitive and behavioral strategies to improve their focus and attention. The study found that children who received CBT showed significant improvements in attention and overall functioning compared to those who received a control intervention.

4. Yoga-Based Relaxation Techniques in Older Adults with Cognitive Impairment

A study published in the Journal of Applied Gerontology explored the effectiveness of yoga-based relaxation techniques in older adults with cognitive impairment. The intervention involved teaching participants yoga poses and breathing techniques designed to promote relaxation and reduce stress. The study found that the intervention significantly improved participants' concentration, working memory, and overall cognitive function.

5. Progressive Muscle Relaxation (PMR) in Adolescents with Anxiety

A study published in the Journal of Psychiatric and Mental Health Nursing evaluated the effectiveness of PMR in reducing anxiety and improving concentration in adolescents. The intervention involved teaching participants to systematically tense and relax their muscles while focusing on their breathing. The study found that after receiving PMR training, participants showed significant reductions in anxiety and improvements in concentration.

Chapter 14

How to create a personalized concentration improvement plan

1. Understand your concentration weaknesses
Begin by **assessing** the areas where you struggle with concentration. For instance, are you easily distracted by notifications or interruptions? Do you find it challenging to stay motivated and engaged throughout the day? Understanding your **weaknesses** will help you develop a plan to address them.
2. Set specific concentration goals
What specific concentration goals do you have for yourself? Be specific and **measurable**: for example, "I want to be able to work on a task for 45 minutes without getting distracted." This will help you stay focused and motivated.
3. Identify what works for you
Different strategies work for different people. Identify **what works best for you** when it comes to concentration improvement. For instance, using apps to block distractions, meditating, breaking down tasks into small chunks, or taking frequent breaks to recharge.
4. Prioritize self-care
Maintaining **good health habits** is essential for improving concentration. Ensure you get enough sleep, eat nutritious meals, exercise regularly, and take breaks when you need to avoid burnout.
5. Start with small steps

Implement your chosen strategies in small, **achievable steps**. For example, commit to taking a five-minute break every hour or using productivity apps for 20 minutes every day. Gradually increase the time spent on these activities until you achieve your concentration goals.

6. Monitor your progress

Track your progress regularly to ensure you **stay on track**. This will help you see the progress you are making, learn from what is working, and refine your approach over time.

By following these steps, you can create a personalized concentration improvement plan that works best for your needs and lifestyle.

The benefits of consistent practice of concentration-boosting activities

There are several benefits of consistent practice of concentration-boosting activities, including:

1. Better focus and attention: Consistent practice of concentration-boosting activities such as deep breathing exercises can improve your concentration and focus. By **training** your brain to focus on a single task, you become better at ignoring distractions and maintaining your attention on what's important.
2. Improved memory: Concentration-boosting activities can also improve your memory by strengthening the neural connections between brain cells. When you're able to focus better, you're more likely to **retain and recall** information more effectively.
3. Reduced stress and anxiety: Consciously focusing your attention on one thing can help to calm your mind and **reduce stress and anxiety**. Regular practice of concentration-boosting activities can help to center your thoughts and bring you a sense of inner calm.
4. Increased creativity: When you are able to concentrate better, you also open up channels for **creative thinking**. Instead of being bogged down by mental chatter, you can access your imagination and inspiration more easily.

5. Improved productivity: Consistent practice of concentration-boosting activities can also lead to improved productivity. By reducing distractions and staying focused on the task at hand, you are more likely to complete your work **efficiently** and **effectively**.

Chapter 15

Frequently Asked Questions & Answers

What is concentration?

Concentration is defined as the ability to focus one's attention on a specific task, object, or thought while excluding distractions. It is a cognitive process that involves our brain's ability to filter and prioritize incoming information based on its relevance.

The Science Behind Concentration:

Our brain has a limited attention span, which can be compared to a muscle that needs to be exercised regularly to stay strong. When we concentrate, our brain allocates resources and energy to the task at hand, allowing us to process information at a deeper level and make better decisions.

Concentration is important for various aspects of our lives, including productivity, creativity, and mental well-being. By improving our ability to focus, we can enhance our learning, memory, and problem-solving skills, and ultimately achieve our goals more efficiently.

Common Distractions That Hinder Concentration:

In today's fast-paced world, we are constantly bombarded with various distractions, such as social media notifications, emails, phone calls, and conversations. These interruptions not only take away our attention from the task at hand but also cause mental fatigue, stress, and anxiety.

To minimize the impact of distractions, it's essential to create a conducive environment that fosters concentration. This could include setting aside dedicated time slots for work, turning off notifications, and practicing mindfulness techniques to reduce stress and anxiety.

Techniques and Strategies for Improving Concentration:
The good news is that concentration is a learnable skill that can be improved with practice and patience. Some effective techniques and strategies for enhancing concentration include the Pomodoro technique (breaking down tasks into smaller, manageable chunks), taking breaks, prioritizing important tasks, and creating a distraction-free environment.
Myth-Busting: Separating Fact from Fiction When it Comes to Concentration and Productivity:
There are several myths and misconceptions surrounding concentration and productivity, such as the idea that multitasking is a time-saver, or that working longer hours equates to being more productive. However, research has shown that these beliefs are not entirely true and can, in fact, hinder performance.
To be truly productive and efficient, it's important to understand the science behind concentration and use evidence-based strategies that work for our individual needs and preferences. By doing so, we can unlock our full potential and achieve our goals with ease.

Why is concentration important?

Concentration is a vital aspect of life that is often overlooked by many. Whether you are a student, an entrepreneur, or an employee, it is crucial to have good concentration skills to achieve success in any field. This is because improved concentration leads to better focus, better decision-making, and increased productivity, which can directly impact your quality of life.
The benefits of improved concentration skills:
Improving your concentration skills can bring about several benefits, including the following:

1. Boosted productivity: With better concentration, you can get more done in less time. This is because you are better able to focus on the task at hand, avoiding distractions that would otherwise eat up your time.
2. Improved memory: Concentration is linked to the hippocampus, a part of the brain responsible for memory formation. By improving your

concentration skills, you are better able to retain important information and recall it when needed.

3. Enhanced creativity: When you can concentrate better, you can generate new ideas and solutions to problems more easily. This is because you can focus on the task at hand without being distracted by irrelevant thoughts.

Common misconceptions about concentration:

Here are some common misconceptions about concentration that you should be aware of:

1. Concentration is an innate skill: While some people may have a natural ability to concentrate better than others, concentration is a skill that can be developed and improved over time with practice.

2. Concentration requires complete silence: Some people assume that to concentrate, they need to be in a completely silent environment. While this may be true for some, others may find that having some background noise helps them focus better.

Strategies for improving concentration skills:

Here are some strategies you can incorporate into your daily routine to improve your concentration skills:

1. Practice mindfulness: Mindful meditation is a popular technique that involves focusing on the present moment without judgment. It can help improve concentration skills by training the mind to focus on one thing at a time.

2. Take breaks: Taking regular breaks throughout the day can help refresh your mind, reduce mental fatigue, and improve concentration.

3. Eliminate distractions: Identify common distractions in your environment and eliminate them as much as possible. For instance, turn off your phone or avoid checking your email during work hours.

How to recognize and address concentration difficulties:

If you are having difficulty concentrating, here are some signs to look out for:

1. Lack of focus and easily distracted: If you find yourself getting easily distracted by irrelevant thoughts or external stimuli, it may be a sign of poor concentration skills.
2. Difficulty completing tasks: If you start several tasks but struggle to finish any of them, it may be due to a lack of concentration.

To address concentration difficulties, consider seeking professional help. You could consult a cognitive behavioral therapist, who can help you identify and address any underlying issues affecting your concentration skills.

What are the common problems with concentration?

Have you ever been sitting at your desk, trying to focus on a task, and found your mind wandering off to something else entirely? Or perhaps you've been asked to read a book or complete a project, only to find yourself getting constantly sidetracked by notifications popping up on your phone or computer.

The truth is, maintaining concentration can be a real challenge. Distractions abound in our modern world, and it can be difficult to stay on task when there are so many tempting diversions vying for our attention.

One of the biggest culprits when it comes to distracting us from our work is technology. With smartphones, tablets, and laptops always within easy reach, it's all too easy to get sucked into checking emails, browsing social media, or responding to text messages. And while multitasking might seem like a good way to get things done more quickly, it tends to be less effective than working on one task at a time.

Another factor that can affect our ability to concentrate is lack of sleep. When we don't get enough rest, our brains aren't able to function as effectively as they should, making it harder to stay focused and productive. Additionally, poor diet and lack of exercise can also impact our concentration levels, making it crucial to prioritize healthy habits to perform at our best.

So, what can we do to improve our concentration and combat distractions? For starters, it's important to take regular breaks throughout the day to give our brains a chance to recharge. Setting goals and practicing mindfulness can also be effective strategies for maintaining focus over long periods of time.

When it comes to minimizing distractions, there are a few practical tips that can be helpful. For example, turning off notifications on our devices or using apps that block certain websites or apps during designated work times. Additionally, it can be helpful to work in a quiet environment or use noise-canceling headphones to minimize external distractions.

Finally, it's worth noting that concentration and mental health are closely linked. Issues like anxiety and depression can significantly impact our ability to focus and seeking help if we're struggling is an important step toward improving our concentration.

Ultimately, our mindset plays a big role in our ability to concentrate effectively. By approaching tasks with a positive attitude, setting manageable goals, and avoiding negative self-talk, we can improve our ability to focus and achieve our goals. So, if you find yourself struggling with concentration, remember that there are strategies and tools available to help you stay on track and reach your full potential.

How can you improve your concentration?

Improving your concentration is essential for success and is a skill that can be sharpened. Increased concentration levels not only enhance your ability to stay focused but also bring along several benefits that can make a significant difference in your life. From productivity to memory retention, the advantages of improved concentration are countless.

One of the most significant benefits of better concentration is increased productivity and efficiency. When you can stay focused on tasks, you can accomplish them more quickly and accurately. This, in turn, allows you to complete more work in less time and with higher quality. As a result, you are more likely to meet your goals and achieve success.

Another advantage of improved concentration is enhanced learning and memory retention. When you can focus on the learning material without getting distracted, you can retain more information and recall it more easily later. Improved concentration levels also allow you to learn new skills more quickly and efficiently.

Additionally, improved concentration levels can reduce stress and anxiety. When you are not distracted by irrelevant thoughts or outside interruptions, you are more likely to remain calm and focused. This reduces stress levels and improves overall well-being.

While there are several rewards for improved concentration levels, many people struggle with poor concentration. Common causes of poor concentration include distractions and interruptions, lack of sleep and rest, and stress and anxiety. Fortunately, several techniques can help improve concentration.

One technique for improving concentration is mindfulness and meditation. Practicing mindful breathing or meditation for a few minutes each day can improve attention span and reduce the anxiety that can distract focus.

Prioritizing tasks and eliminating distractions can also help maintain focus. Use time management skills like the Pomodoro method or apps to limit distractors among notifications, media, and other sources.

Furthermore, regular exercise and a healthy diet also do wonders to improve concentration. Regular exercises improve cognitive abilities and release stress, and a diet that prioritizes protein and nutrient-rich foods, big leafy greens, less sugar, and caffeine reduces mood swings and improves focus.

While these techniques can be helpful, it's also important to seek professional help when concentration difficulties persist despite efforts to improve. If your concentration issues are affecting daily life and functioning, it's time to get help from a mental health or medical professional.

In conclusion, improving your concentration is beneficial to your daily life and success. You can achieve improved concentration by prioritizing tasks, eliminating distractions, practicing mindfulness, exercising, and having a healthy diet. When concentration issues persist, don't hesitate to seek help from a professional.

What are some exercises to improve concentration?

Are you struggling to focus on work or on your studies? Maybe you find yourself easily distracted and unable to concentrate on the task at hand. The good news is that there are many exercises you can do to help improve your ability to concentrate. Here are just a few:

The Power of Meditation: How to Improve Your Ability to Concentrate

Meditation is an excellent tool for sharpening your concentration skills. By focusing your attention on your breath or a specific object, you can train your mind to stay grounded in the present moment. As you practice, you may find that it becomes easier to stay focused for longer periods of time.

Unplugging for Focus: The Benefits of Digital Detoxing for Your Brain

We live in a world where we're surrounded by electronic devices, and it can be challenging to escape the constant barrage of notifications, emails, and social media. A digital detox can help reset your brain and improve your ability to concentrate. Try turning off your phone and computer for a set period each day and see how it affects your focus.

Sweat It Out: The Role of Exercise in Enhancing Concentration

Exercise is not just essential for physical health; it also has many benefits for the brain. When you exercise, your body releases chemicals called endorphins, which can help improve your mood and reduce stress. Additionally, regular exercise has been shown to increase focus and cognitive function, making it an excellent tool for improving concentration.

Fueling Your Brain: The Best Foods to Eat for Optimal Cognitive Function

Finally, what you eat can play a significant role in your ability to concentrate. Eating a diet rich in whole foods, healthy fats, and complex carbohydrates can help keep your brain functioning optimally. On the other hand, consuming too much sugar or processed food can lead to brain fog and decreased cognitive function.

In conclusion, there are many exercises you can do to help improve your ability to concentrate. Whether it's meditation, exercise, or simply unplugging technology, taking a break from distractions can help sharpen your focus and improve your productivity. So why not try one of these exercises today and see how it affects your concentration?

How can you maintain concentration for longer periods?

We've all been there – sitting down to work on a project, only to find our minds wandering off after just a few minutes. Maintaining concentration for a prolonged period can be a struggle, but it's not impossible. By understanding the psychological factors at play, incorporating mindfulness exercises into our daily routine, maintaining a healthy diet and exercise regimen, and learning to manage our technology use, we can train our brains to stay focused and reduce distractions.

First and foremost, it's important to understand the psychology of concentration. Our brains are wired to get distracted easily, as we are constantly bombarded by stimuli from our environment. However, by understanding this tendency and recognizing when our attention begins to wane, we can take steps to bring our focus back to the task at hand.

One way to do this is by incorporating mindfulness exercises into our daily routine. These exercises help us train our minds to stay focused and reduce distractions, while also promoting relaxation and stress reduction. Examples of mindfulness exercises include deep breathing, meditation, and mindful movement practices like yoga or tai chi.

In addition to mindfulness, our diet and exercise habits can also play a significant role in our ability to concentrate. Eating a nutritious diet rich in brain-boosting foods like omega-3 fatty acids, antioxidants, and complex carbohydrates can help improve focus and concentration. Similarly, participating in regular physical activity can help increase blood flow to the brain, improving cognitive function and mental clarity.

Lastly, our constant exposure to technology can also impact our attention span. While technology can be a helpful tool, it's important to learn to manage our use and incorporate breaks from screens. This can include taking short breaks to rest our eyes or engage in physical activity, setting boundaries around technology use during certain times of day, and finding alternative activities that allow us to unplug and recharge.

In conclusion, maintaining concentration for longer periods of time requires a multifaceted approach that addresses the psychological, physical, and technological factors at play. By incorporating mindfulness exercises into our daily routine, maintaining a healthy diet and exercise regimen, and learning to manage our technology use, we can help train our minds to stay focused and reduce distractions.

How do distractions affect concentration?

Do you find yourself struggling to stay focused on a task? Do you often feel like the world around you are vying for your attention, making it difficult to concentrate and get work done? You're not alone. Concentration is something that many people struggle with, and distractions can often be the culprit. But what exactly is the effect of distractions on concentration?

Distractions can have a significant impact on concentration, leading to decreased productivity and increased errors in work. With the overwhelming amount of technology and information at our fingertips, it can be challenging to keep our focus on one thing. While technology can be an incredibly helpful tool, it can also be a major source of distraction. Think about notifications popping up on your phone or the lure of social media.

So, how do we avoid distractions and improve our concentration? One strategy is to set boundaries. Turn your phone on silent and put it in a different room if possible. Disable notifications on your computer or limit checking your email to specific times of the day. Time management is also key. Try breaking down your work into smaller, more manageable tasks, and take breaks when needed to recharge your brain.

Another strategy is mindfulness. Mindfulness is the practice of being present now and paying attention to your thoughts, feelings, and environment. By practicing mindfulness, you learn to recognize when your mind is wandering and bring your focus back to the task at hand.

But what if distractions are a chronic issue? It's essential to identify and address any underlying issues that may be causing distraction, such as anxiety or ADHD. These conditions can affect concentration and require professional help to manage effectively.

In conclusion, distractions can have a significant impact on concentration, leading to decreased productivity and increased errors in work. But with strategies such as setting boundaries, time management, and mindfulness, we can improve our focus and be more productive. And if distractions are a chronic problem, seeking professional help is key to identifying and managing any underlying issues.

What are some tips for staying focused?

Staying focused can be a daunting task, especially when you have a million things that demand your attention. Whether you're working on a project or studying for an exam, maintaining concentration requires a certain level of discipline and practice.

One of the most important things you can do to stay focused is to minimize distractions. Distractions can come in many forms, from social media notifications to the piles of clutter on your desk. To declutter your workspace, try tidying up and organizing your materials. Limiting interruptions by turning off your phone or email notifications can also go a long way in helping you stay focused.

Another tip for staying focused is to prioritize and break down tasks into smaller, more manageable pieces. This can help prevent overwhelm and increase productivity. Try creating a to-do list and prioritizing tasks by deadline or importance. Breaking down tasks into smaller goals can also help you make progress and see results, which can be a source of motivation.

Taking breaks is also essential for concentration. Research has shown that taking regular breaks can increase productivity and focus. Incorporating mindfulness practices, such as deep breathing or meditation during breaks, can also help you recharge and refocus.

If you find yourself struggling with concentration despite these tips, it may be time to seek additional support or resources. Time management techniques, such as the Pomodoro technique or setting specific work hours, can help you structure your day and make the most of your focus. For some, medication therapy may be necessary to manage attention and concentration issues.

Remember, focus is a skill that can be improved with consistent practice and discipline. Use these tips to create a focused and productive environment, and don't be afraid to seek support if you need it.

How can you use technology to improve concentration?

In today's fast-paced digital world, it can be difficult to concentrate on one task for an extended period. With a constant stream of notifications and distractions, it's no wonder many of us struggle with maintaining focus. But what if we told you that there's a way to harness technology to help improve your concentration? Here are some tips on how to do just that.

Firstly, it's important to acknowledge the impact that digital distractions can have on concentration. The constant notifications from social media, email and messaging apps can interrupt our workflow and make it difficult to focus. That's why it's essential to turn off notifications or place your phone in another room altogether when trying to concentrate on a task.

But technology can also assist with concentration. There are numerous apps and tools available that can help improve productivity and minimize distractions. From time-management apps like Rescue Time to the Pomodoro technique timer, technology can be used to help create a structured and efficient work schedule.

However, it's important to be cautious when using technology to enhance concentration. Excessive use of productivity apps or digital tools can be counterproductive and lead to burnout. It's important to take breaks from technology and allow ourselves time to recharge and refocus.

One way to use technology in a productive and focused manner is to set designated periods of time for work, followed by a break. During this break, consider taking a walk, practicing mindfulness, or engaging in a non-digital hobby. This will allow your mind to rest and recharge before returning to work, ultimately improving your concentration and productivity in the long term.

In conclusion, it's important to use technology in a mindful and strategic way to improve concentration. While it can be a useful tool, it's equally essential to take breaks and prioritize rest and self-care. By doing so, you'll be able to enhance your focus and create a more productive and fulfilling work routine.

In conclusion, concentration is the key to success, and it is achievable with mindful attention and focus. Whether you are a student or a professional, practicing concentration can help you enhance your productivity and overall well-being. By implementing the tips and techniques shared in this article, you can become more focused, responsive, and productive in your daily life. Let's make concentration a habit and reap the benefits it has to offer! Happy focusing!

How long should I sit and study without taking a break?

It is generally recommended to take a short break after every 25-30 minutes of focused study. This includes taking a few minutes to stand up, stretch, and move around, to give your brain a chance to rest and recharge. Continuous studying without taking breaks can lead to exhaustion and decreased productivity. It is essential to listen to your body's needs and take breaks as needed to maintain concentration and focus.

Are there any specific foods or supplements that can help improve concentration?

Some foods and supplements that might help improve concentration include:

1. Omega-3 fatty acids: These essential fatty acids, found in fish such as salmon, tuna, and sardines, can improve brain function and learning ability.
2. Blueberries: These berries are rich in antioxidants and flavonoids that can help protect the brain from oxidative stress and improve cognitive function.
3. Dark chocolate: It contains flavonoids that can enhance blood flow to the brain and improve cognitive function.
4. Green tea: It contains an amino acid called L-theanine, which can increase alpha wave activity in the brain and improve focus and attention.
5. Caffeine: It can improve alertness and concentration, but it should be consumed moderately.

6. Ginseng: It can improve mental performance, including concentration, memory, and reaction time.

However, it's essential to note that a healthy and balanced diet, regular exercise, and adequate sleep are crucial for maintaining optimal cognitive function. Also, consult a healthcare professional before taking any new supplements.

What are some effective study techniques for maintaining focus and concentration?

1. Create a distraction-free study environment. This means finding a quiet and comfortable area where you won't be interrupted.
2. Break your study sessions into smaller chunks. Focus on one task or topic for a set amount of time (like 20-30 minutes) then take a short break. Repeat for several hours a day.
3. Use a timer or Pomodoro technique. Use a timer to stay on track and focused by taking short breaks every 25 minutes.
4. Prioritize assignments and create a to-do list. Organize your study sessions by prioritizing what needs to be done first.
5. Review notes often. Reviewing course notes on a regular basis can keep the material fresh in your mind.
6. Try incorporating active learning techniques, such as taking notes, answering questions, testing yourself, and engaging in discussions.
7. Designate certain study times. Establish a schedule and follow it. This can help your mind and body get into a routine.
8. Reduce distractions caused by technology. Turn off notifications, notifications, mute your phone or put it in a different room.
9. Get enough sleep and exercise. Reduce stress by sleeping and exercising regularly.
10. Drink enough water and eat healthy snacks. Your body and brain operate best when you are hydrated and eating healthy snacks.

Remember, it's important to tailor your study techniques to your way of learning, so try different methods to see what is most effective for you.

Are there any particular exercises or activities that can help improve concentration?

Yes, there are several exercises and activities that can help improve concentration. Here are some of the most effective ones:

1. Mindfulness Meditation: Mindfulness meditation is a technique that involves focusing on the present moment without judgment. It has been found to be an effective way to improve concentration and focus.
2. Regular Exercise: Regular exercise has been found to have a positive effect on cognitive function, including concentration. Exercise also helps reduce stress and anxiety, which can be major distractions.
3. Brain Games: There are many brain games, such as crossword puzzles, Sudoku, and memory games that are designed to improve concentration and cognitive function.
4. Deep Breathing: Deep breathing exercises can help you calm your mind and improve focus. They help increase the oxygen supply to the brain, which helps in improving concentration.
5. Reading: Reading is an activity that requires a high level of concentration. It helps improve focus and attention span over time.
6. Mind Mapping: Mind mapping is a technique that involves creating a visual map of ideas, concepts, or information. This helps improve concentration, memory, and recall.
7. Listening to Music: Listening to music has been found to improve concentration and cognitive function. Some types of music, such as instrumental or classical music, are particularly effective in enhancing focus.
8. Taking Breaks: Taking regular breaks during work or study sessions can help improve concentration. Experts recommend taking a 5–10-minute break every hour to refresh your mind and prevent burnout.

How can I reduce distractions while studying?

1. Find a quiet place to study: Avoid studying in noisy or crowded places, such as busy cafes or public areas. Find a quiet place, such as a library or a study room, where you won't be disturbed.
2. Turn off electronic devices: Turn off your cell phone or set it to airplane mode to avoid interruptions from notifications, text messages, or calls. Also, consider disabling notifications on your computer.
3. Use a timer: Set a timer for a specific period of time, such as 30 minutes, and focus entirely on your studies during that time. Once the timer goes off, take a break to recharge your brain.
4. Wear headphones/earplugs: Noise-cancelling headphones or earplugs can help block out distracting sounds and help you focus.
5. Create a to-do list: Write down a list of tasks that you want to accomplish during your study session. This will help you stay on track and minimize distractions.
6. Organize your study area: Keep your study area clean and tidy. Remove any items that may cause distractions, such as unnecessary gadgets, books, or decorations.
7. Take breaks: Taking short breaks after a certain period of study will help you recharge and prepare for the next study session.

By implementing these tips, you can minimize distractions and improve your focus while studying.

How long should I sit and study without taking a break?

It is generally recommended to take a short break after every 25-30 minutes of focused study. This includes taking a few minutes to stand up, stretch, and move around, to give your brain a chance to rest and recharge. Continuous studying without taking breaks can lead to exhaustion and decreased productivity. It is essential to listen to your body's needs and take breaks as needed to maintain concentration and focus.

Are there any specific foods or supplements that can help improve concentration?

Some foods and supplements that might help improve concentration include:

1. Omega-3 fatty acids: These essential fatty acids, found in fish such as salmon, tuna, and sardines, can improve brain function and learning ability.
2. Blueberries: These berries are rich in antioxidants and flavonoids that can help protect the brain from oxidative stress and improve cognitive function.
3. Dark chocolate: It contains flavonoids that can enhance blood flow to the brain and improve cognitive function.
4. Green tea: It contains an amino acid called L-theanine, which can increase alpha wave activity in the brain and improve focus and attention.
5. Caffeine: It can improve alertness and concentration, but it should be consumed moderately.
6. Ginseng: It can improve mental performance, including concentration, memory, and reaction time.

However, it's essential to note that a healthy and balanced diet, regular exercise, and adequate sleep are crucial for maintaining optimal cognitive function. Also, consult a healthcare professional before taking any new supplements.

What are some effective study techniques for maintaining focus and concentration?

1. Create a distraction-free study environment. This means finding a quiet and comfortable area where you won't be interrupted.
2. Break your study sessions into smaller chunks. Focus on one task or topic for a set amount of time (like 20-30 minutes) then take a short break. Repeat for several hours a day.
3. Use a timer or Pomodoro technique. Use a timer to stay on track and focused by taking short breaks every 25 minutes.
4. Prioritize assignments and create a to-do list. Organize your study sessions by prioritizing what needs to be done first.

5. Review notes often. Reviewing course notes on a regular basis can keep the material fresh in your mind.
6. Try incorporating active learning techniques, such as taking notes, answering questions, testing yourself, and engaging in discussions.
7. Designate certain study times. Establish a schedule and follow it. This can help your mind and body get into a routine.
8. Reduce distractions caused by technology. Turn off notifications, notifications, mute your phone or put it in a different room.
9. Get enough sleep and exercise. Reduce stress by sleeping and exercising regularly.
10. Drink enough water and eat healthy snacks. Your body and brain operate best when you are hydrated and eating healthy snacks.

Remember, it's important to tailor your study techniques to your way of learning, so try different methods to see what is most effective for you.

Are there any particular exercises or activities that can help improve concentration?

Yes, there are several exercises and activities that can help improve concentration. Here are some of the most effective ones:

1. Mindfulness Meditation: Mindfulness meditation is a technique that involves focusing on the present moment without judgment. It has been found to be an effective way to improve concentration and focus.
2. Regular Exercise: Regular exercise has been found to have a positive effect on cognitive function, including concentration. Exercise also helps reduce stress and anxiety, which can be major distractions.
3. Brain Games: There are many brain games, such as crossword puzzles, Sudoku, and memory games that are designed to improve concentration and cognitive function.
4. Deep Breathing: Deep breathing exercises can help you calm your mind and improve focus. They help increase the oxygen supply to the brain, which helps in improving concentration.
5. Reading: Reading is an activity that requires a high level of concentration. It helps improve focus and attention span over time.

6. Mind Mapping: Mind mapping is a technique that involves creating a visual map of ideas, concepts, or information. This helps improve concentration, memory, and recall.
7. Listening to Music: Listening to music has been found to improve concentration and cognitive function. Some types of music, such as instrumental or classical music, are particularly effective in enhancing focus.
8. Taking Breaks: Taking regular breaks during work or study sessions can help improve concentration. Experts recommend taking a 5–10-minute break every hour to refresh your mind and prevent burnout.

Chapter 16

Resources

Additional reading

1. "Deep Work: Rules for Focused Success in a Distracted World" by Cal Newport
2. "The Power of Concentration" by Theron Q. Dumont
3. "Flow: The Psychology of Optimal Experience" by Mihaly Csikszentmihalyi
4. "The One Thing: The Surprisingly Simple Truth Behind Extraordinary Results" by Gary W. Keller and Jay Papasan
5. "The Willpower Instinct: How Self-Control Works, Why It Matters, and What You Can Do to Get More of It" by Kelly McGonigal
6. "How to Focus Better: Practical Tips for Improving Your Concentration" by James Clear
7. "Peak Performance: Elevate Your Game, Avoid Burnout, and Thrive with the New Science of Success" by Brad Stulberg and Steve Magness
8. "Atomic Habits: An Easy & Proven Way to Build Good Habits & Break Bad Ones" by James Clear
9. "The Art of Possibility: Transforming Professional and Personal Life" by Rosamund Stone Zander and Benjamin Zander
10. "Make Time: How to Focus on What Matters Every Day" by Jake Knapp and John Zeratsky.

Chapter 17

App recommendations for time management and productivity

1. Trello
2. Asana
3. Todoist
4. StayFocusd
5. RescueTime
6. Focus@Will
7. Forest
8. Toggl
9. Pomodoro Timer
10. Google Calendar

Chapter 18

Online resources and courses on concentration improvement.

1. Elevate - Brain Training: This app offers various fun and challenging games that help improve memory, focus, and analytical skills.
2. Headspace: A popular meditation app that provides guided meditations to help improve focus and concentration.
3. Focus@Will: This app offers music tracks specifically designed to help improve focus, reduce distractions, and increase productivity.
4. Lumosity: Provides a wide range of brain-training games to work on different skills like attention, processing speed, and memory retention.
5. Calm: Another popular meditation app that helps reduce stress and improve focus.
6. Udemy: Offers various online courses on improving concentration and focus.
7. Coursera: Provides access to online courses from top universities on the topics of cognitive psychology, neuroscience, and attention improvement.
8. Khan Academy: Offers free video lessons and interactive resources on how to improve focus and study skills.
9. Skillshare: Provides online courses on productivity and concentration improvement offered by experts in the field.
10. Mindvalley: Offers personal growth courses that help improve focus and concentration as well as mindfulness practices.

Chapter 19

Conclusion

To recall the different strategies for Top Techniques to Improve your Concentration:

1. Prioritize and schedule tasks,
2. Exercise regularly,
3. Get enough sleep,
4. Practice deep breathing exercises,
5. Take breaks when needed,
6. Limit distractions,
7. Reduce multitasking,
8. Listen to music,
9. Use positive affirmations.

By following these techniques, you can improve your concentration and be more productive in your daily life.

To summarize the main points or tips provided throughout this book, consider these:

1) Understanding the importance of concentration for achieving success in various areas of life, including work, education, and personal goals.
2) Recognizing and avoiding distractions, such as social media, multitasking, and internal thoughts.
3) Practicing breathing exercises to calm the mind and increase focus.
4) Incorporating physical exercise and a healthy diet into daily routines to improve overall cognitive function.

5) Creating a structured schedule and prioritizing tasks to reduce stress and improve productivity.

Overall, the book might emphasize that improved concentration is not only beneficial to achieving short-term goals but also to increase overall well-being and fulfillment in life.

Good to have you here!

Scan the below QR code to more great books

Get more beneficial books at:

https://booksforminds.com/[1]

1. https://gotowebsite.online/ebkconcent

www.ingramcontent.com/pod-product-compliance
Lightning Source LLC
LaVergne TN
LVHW022310200926
848481LV00004B/1268
* 9 7 9 8 2 1 5 5 2 9 4 8 5 *